Executive Summary

# the costs of the israeli-palestinian conflict

This summary volume is based on the comprehensive report
*The Costs of the Israeli-Palestinian Conflict*

C. Ross Anthony
Daniel Egel
Charles P. Ries
Mary E. Vaiana

For more information on this publication, visit www.rand.org/t/rr740z1-1

This revised edition incorporates minor editorial changes and includes some additional data from the comprehensive report, *The Costs of the Israeli-Palestinian Conflict.*

## The Costs-of-Conflict Study Team

C. Ross Anthony, Daniel Egel, Charles P. Ries
Craig A. Bond, Andrew M. Liepman, Jeffrey Martini
Steven Simon, Shira Efron, Bradley D. Stein
Lynsay Ayer, Mary E. Vaiana

**Library of Congress Cataloging-in-Publication Data** is available for this publication.
ISBN: 978-0-8330-9034-8

Published by the RAND Corporation, Santa Monica, Calif.


*Cover design: Sean Christensen and Doug Suisman*

To David K. Richards

His courage, vision, and humanity inspired all of us
who had the great privilege of knowing him.

# Maps

**Map 1**
**Israel, the West Bank, and Gaza**

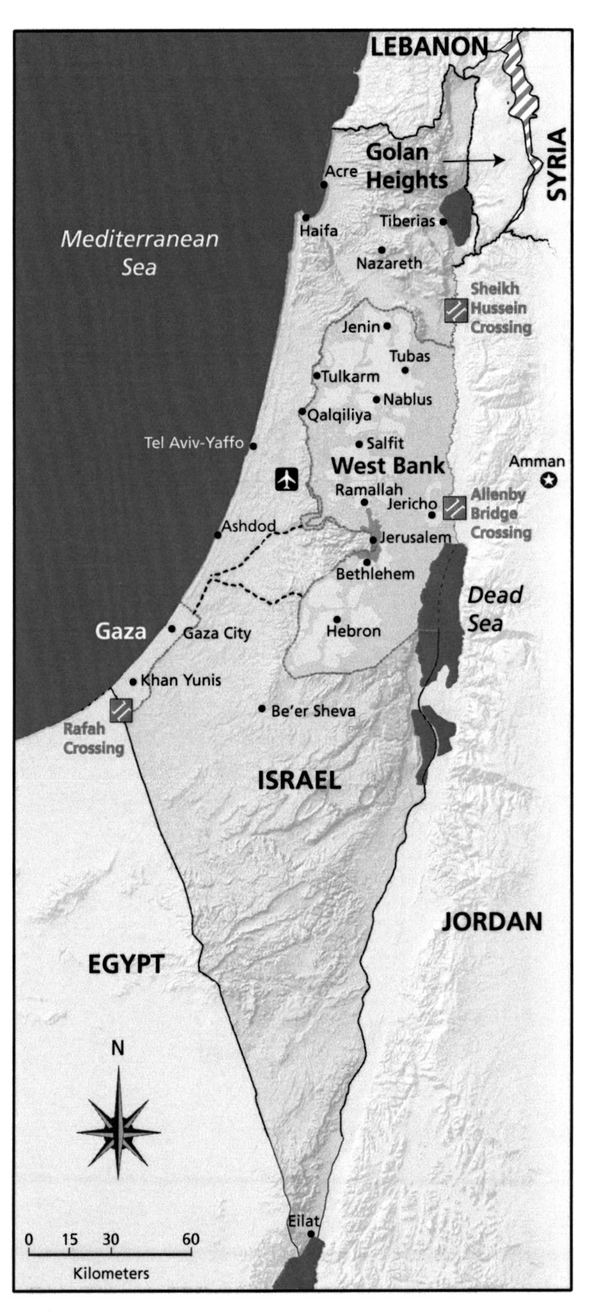


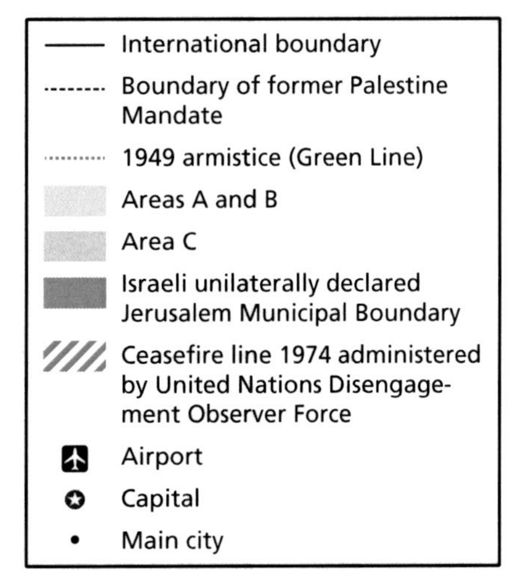


SOURCE: Adapted from United Nations Office for the Coordination of Humanitarian Affairs (OCHA) Occupied Palestinian Territory (oPt) map.

RAND *RR740/1-1-Map1*

**Map 2**
**Israeli Settlements in the West Bank, January 2012**

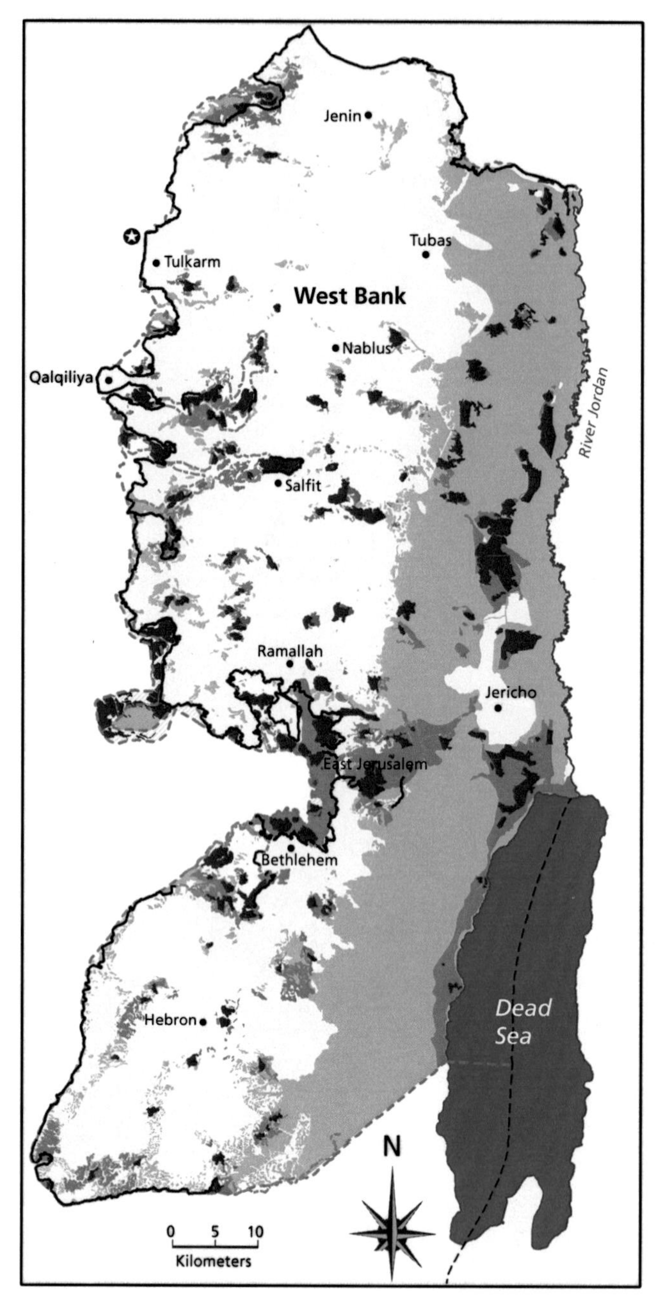


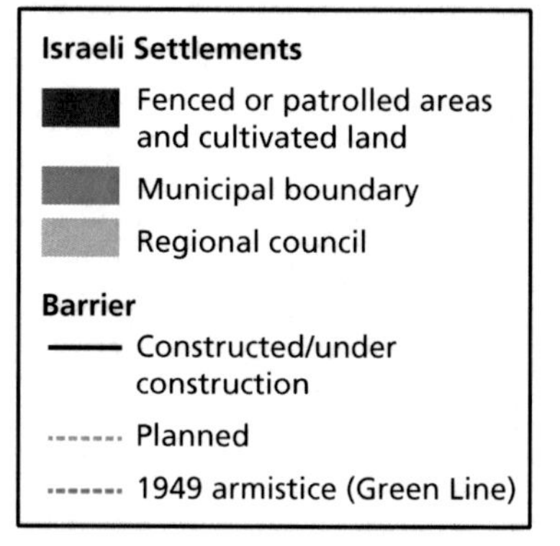


SOURCE: Adapted from United Nations OCHA oPt map.
RAND *RR740/1-1-Map2*

# Preface

This executive summary highlights main findings from the complete report *The Costs of the Israeli-Palestinian Conflict*. The comprehensive report and all related study materials are available on the project website at www.rand.org/costsofconflict.

The study estimates the net costs and benefits if the long-standing conflict between Israelis and Palestinians follows its current trajectory over the next ten years, relative to five other possible trajectories that the conflict could take. The goal of the analysis is to give all parties comprehensive, reliable information about available choices and their expected costs and consequences.

As the regional context for this work is dynamic, we had to choose a date on which to cut off data collection. Data for this analysis were collected from February 2013 through April 2014. Since April 2014, we have tried to incorporate information about certain key events in our discussion, but we have not integrated that information in any systematic way, and it is not reflected in our analytic findings.

Economic and political assumptions underpinning the analysis drive important outcomes. We have clearly specified those assumptions in our discussion. However, the project website also houses a costing tool (www.rand.org/cc-calculator), making it possible for users to change key analytic assumptions and explore how the changes affect outcomes.

This work should be of interest to policymakers in Israel, the West Bank and Gaza, and the Middle East more generally; to the international community; to foreign policy experts; and to organizations and individuals committed to finding a permanent and peaceful resolution to the conflict.

The study was supported by a generous gift from David and Carol Richards.

Questions or comments about the work should be sent to the project leaders, C. Ross Anthony (rossa@rand.org) or Charles P. Ries (ries@rand.org).

## The Center for Middle East Public Policy

The research described in this report was conducted within the Center for Middle East Public Policy (CMEPP), part of International Programs at the RAND Corporation. CMEPP brings together analytic excellence and regional expertise from across the

RAND Corporation to address the most critical political, social, and economic challenges facing the Middle East today. For more information about the RAND Center for Middle East Public Policy, visit http://www.rand.org/cmepp.html or contact the center director (contact information is provided on the center's web page).

# Acknowledgments

Many knowledgeable experts on the conflict, including more than 200 Israelis, Palestinians, Europeans, and Americans, took time from their busy schedules to meet with members of the RAND research team, some repeatedly, as we defined our research objectives and worked through the challenges of devising objective assumptions for present trends projections and our analytical scenarios. We thank them for their patience and wisdom.

We also thank Warren Bass and David Schoenbaum for their work on a concise and balanced history of the conflict from 1967 to the present. While we ultimately decided not to include the chapter in our research report since there is already an extensive literature on the subject, their "history of the conflict" discussion played a key role in informing our research approach and in ensuring that all team members had a firm grounding in the conflict's history over the past half century.

Any project like this should be firmly anchored in reality and stress tested by those who live the trends it purports to analyze. From the outset, therefore, we intended to lay out the evolving research design in a workshop involving Israeli, Palestinian, and European experts with decades of experience and insight. The workshop, held in Athens, Greece, in April 2014, was hosted by the government of Greece and benefited from suggestions, course corrections, and insights of nearly 20 top experts. The research team wishes to thank the Greek Ministry of Foreign Affairs, then-Minister Evangelos Venizelos, and Secretary General for International Economic Cooperation Peter Mihalos for their extraordinary hospitality and support.

We are especially grateful to our technical reviewers, whose wisdom and patience helped to sharpen our analysis and clarify our presentation. We thank Arie Arnon, Keith Crane, David Johnson, Daniel Kurtzer, Karim Nashashibi, Marc Otte, and James Quinlivan for their insights. In addition to his role as a formal reviewer, Keith Crane functioned as an embedded adviser, helping the team address difficult analytic issues and refine critical assumptions. Shanthi Nataraj provided constructive comments on the costing tool, ensuring its accuracy and accessibility. We also thank Shlomo Brom, Hiba Husseini, Dalia Dassa Kaye, Carol Richards, Michael Schoenbaum, Claire Spencer, and Jeffrey Wasserman for informal comments along the way. Shmuel

Abramzon, Sean Mann, and Semira Yousefzadeh provided vital research assistance. Of course, the authors assume full responsibility for the study findings.

Many members of RAND's Office of External Affairs contributed to this document. In particular, we thank Jocelyn Lofstrom for coordinating all aspects of production, including translations and the project website; Sandy Petitjean for her sustained efforts to produce high-quality graphics; and Nora Spiering, whose careful editing greatly improved the clarity of our presentation. We are also grateful for her patience in assembling an integrated draft of the manuscript on multiple occasions to facilitate our work and for managing and verifying the extensive references. In addition, we thank James Torr, who also did considerable work on references. Ingrid Maples provided invaluable administrative assistance at many stages of the project, and her energy and talent for logistics were vital to the success of the Athens workshop.

We wish to thank Michael Rich, RAND's president and chief executive officer, for his unwavering support of this project and the unbiased, objective analysis it seeks to provide.

We are especially grateful to David and Carol Richards. This study would not have been possible without their generous intellectual and financial support. Their commitment to peace in the Middle East remains a source of inspiration to us all.

# Contents

# Maps, Figures, and Tables

## Maps



## Figures



## Tables

# Executive Summary

For much of the past century, the conflict between Israelis and Palestinians has been a defining feature of the Middle East. Despite billions of dollars expended to support, oppose, or seek to resolve it, the conflict has endured for decades, with periodic violent eruptions, of which the Israel-Gaza confrontation in the summer of 2014 was only the most recent.

This study estimates the net costs and benefits over the next ten years of five alternative trajectories—a **two-state solution**, **coordinated unilateral withdrawal**, **uncoordinated unilateral withdrawal**, **nonviolent resistance**, and **violent uprising**—compared with the costs and benefits of a continuing impasse that evolves in accordance with present trends. The analysis focuses on economic costs related to the conflict, including the economic costs of security. In addition, we calculate the costs of each scenario to the international community. Unless otherwise indicated, all costs are denoted in constant 2014 U.S. dollars.

To the degree possible, we consider intangible factors, such as distrust, religion, and the fear of relinquishing some degree of security, and how such factors might affect future pathways.

The study's focus emerged from an extensive scoping exercise designed to identify how RAND's objective, fact-based approach might promote fruitful policy discussion. We reviewed previous research on key dimensions of the problem and, where possible and necessary, we conducted additional research to clarify and define issues. Our overarching goal is to give all parties comprehensive, reliable information about available choices and their expected costs and consequences.

We integrated findings from our fieldwork, the literature review, and our supplemental analyses. Seven key findings emerge from our work:

- A two-state solution provides by far the best economic outcomes for both Israelis and Palestinians. Israelis gain over two times more than the Palestinians in absolute terms—$123 billion versus $50 billion over ten years.
- But the Palestinians gain more proportionately, with average per capita income increasing by approximately 36 percent over what it would have been in 2024, versus 5 percent for the average Israeli.

- A return to violence would have profoundly negative economic consequences for both Palestinians and Israelis; we estimate that per capita gross domestic product (GDP) would fall by 46 percent in the West Bank and Gaza (WBG) and by 10 percent in Israel by 2024.
- In most scenarios, the value of economic opportunities gained or lost by both parties is much larger than expected changes in direct costs.
- Unilateral withdrawal by Israel from the West Bank imposes large economic costs on Israelis unless Israel coordinates with the Palestinians and the international community, and the international community shoulders a substantial portion of the costs of relocating settlers.
- Intangible factors, such as each party's security and sovereignty aspirations, are critical considerations in understanding and resolving the impasse.
- Taking advantage of the economic opportunities of a two-state solution would require substantial investments from the public and private sectors of the international community and from both parties.

## Approach

Understanding the costs of the political impasse requires a methodology to compare conditions that Israelis and Palestinians experience today with what conditions might be under alternative assumptions about political conditions. We use a counterfactual approach, which allows us to explore systematically how specific outcomes might have differed if conditions had been different (see Figure S.1).

**Figure S.1**
**Framework for Analysis**

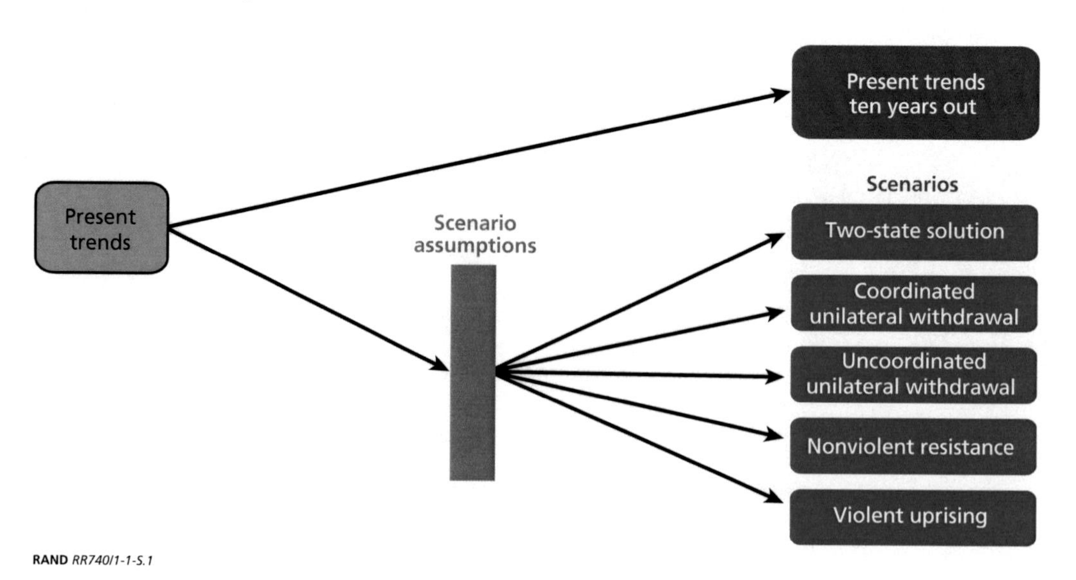

Our base case, which we refer to as "present trends," assumes that economic and security outcomes continue along their current trajectories—i.e., the final status accord issues defined in the Oslo Accords remain unresolved, and there are no significant shocks or changes to economic, demographic, and security conditions. We assume that the impasse remains dynamic, as it has always been, and that conditions, including periodic business disruptions, flare-ups of military engagement, and continued construction of Israeli settlements, continue to evolve along current trajectories.

We use present trends as a baseline reference and compare outcomes, such as GDP or perceptions of security risk, under that reference case to those under five alternative trajectories:

- **A two-state solution**, in which a sovereign Palestinian state is established alongside Israel. Our two-state solution scenario is based on an amalgam of the Clinton Parameters, the Olmert-Abbas package, and the track-two Geneva Initiative.
- **Coordinated unilateral withdrawal** by Israel from a good portion of the West Bank. Our assumptions are based on the work of Israeli nongovernmental institutions, including Blue White Future and the Institute for National Security Studies. The scenario assumes Israeli coordination with both the Palestinians and the international community.
- **Uncoordinated unilateral withdrawal** by Israel from part of the West Bank. We modify assumptions for coordinated unilateral withdrawal to reflect the situation in which Palestinians and the international community do not support Israel's actions or coordinate with it.
- **Nonviolent resistance** by Palestinians in pursuit of their national aspirations. In this scenario, we consider Palestinian legal efforts at the United Nations (UN) and other world bodies, continued support for trade restrictions on Israel, and nonviolent demonstrations.
- **Violent uprising.** In this scenario, we consider the effects of a violent Palestinian uprising, perhaps emanating from Gaza but also including the West Bank and possibly participation from foreign terrorists.

For each scenario, we derive economic and security assumptions based on historical precedent, a review of the existing literature, and conversations with subject matter experts. The scenarios themselves were designed with four core criteria in mind: They should be credible, they must be sufficiently distinct from other scenarios to warrant analysis, they must be feasible in the ten-year time frame for the analysis, and our counterfactual approach must be appropriate for them.

*We use these scenarios as possible alternative futures for analytic purposes, but we make* ***no prediction*** *about the likelihood of any of them becoming reality.* Indeed, the reality that evolves is likely to be a mixture of some aspects of all the scenarios presented here. For example, the Gaza war in the summer of 2014, the subsequent recriminations, the

Palestinian diplomatic moves at the UN and its agencies, and the punitive responses from Israel depart from what we originally defined as present trends and seem more akin to our nonviolent resistance scenario.

To avoid ambiguity, we attempt to use terminology precisely. We use the term *Israel* to refer to the State of Israel and to the territory defined by the Green Line. We use *Israelis* to designate Israel's inhabitants in general and *Jewish Israelis* and *Palestinian citizens and residents of Israel* when appropriate to distinguish between these groups. We use the term *Palestine* or *West Bank and Gaza* (*WBG*) to refer to the area of the West Bank and the Gaza Strip as defined by the Green Line. We use *State of Palestine* when appropriate to refer to such a prospective entity, especially with respect to our two-state solution scenario, and *Palestinian Authority* (PA) to refer to the entity set up after the Oslo Accords to administer parts of the West Bank and Gaza. We use the term *Palestinians* to refer to the inhabitants (except settlers) of the West Bank, Gaza, and East Jerusalem.

The counterfactual approach has several limitations. First, the results are driven by and are sensitive to the assumptions. The scenarios we use rely on historical precedent when possible, but we also base assumptions on other available evidence and on discussions with subject matter experts. Second, with certain important exceptions, we conduct this research assuming that changes across each of these different dimensions are essentially independent of the others. Such simplification allows us to sum the results of each change; however, it does not accurately reflect real-world linkages locally or regionally.

We have limited our analysis to ten years—effectively, 2014 through 2024. Both positive and negative effects under each scenario can change dramatically in the long run; thus, our estimates should be considered lower bounds.

## Defining Present Trends

The political impasse affects Israelis and Palestinians primarily through its impact on the economic, security, and sociopsychological components of their lives.

### Economics

Our analysis of economic performance focuses on GDP, GDP per capita, and public and private expenditures.

*Gross Domestic Product.* We use historical GDP growth rates to project how the economies of Israel, the WBG, and East Jerusalem are likely to evolve over the next ten years if present trends continue (Table S.1). Because we believe that more recent growth rates are more likely to provide accurate estimates, we use growth rates from 1999 through 2013. This period includes the economic downturn of the Second Intifada (2000–2005), the economic recovery following the Second Intifada, and the economic side effects of Hamas' rise to power in Gaza and several Israeli military operations.

**Table S.1**
**Ten-Year Projections of GDP and GDP Per Capita Under Present Trends: Israel Compared with the West Bank, Gaza, and East Jerusalem**

| | Israel | | | West Bank, Gaza, and East Jerusalem | | |
|---|---|---|---|---|---|---|
| | Average Growth Rate (1999–2013) | 2014 | 2024 | Average Growth Rate (1999–2013) | 2014 | 2024 |
| GDP (U.S.$, billions) | 4.1% | $295 | $439 | 3.6% | $13.9 | $19.9 |
| GDP per capita | 1.9% | $35,900 | $43,300 | 0.6% | $2,890 | $3,080 |

NOTE: Data have been rounded.

The numbers in Table S.1 do not reflect any evolving long-term impact of the July 2014 conflict between Israel and Gaza, which had significant short-term negative effects on both the Israeli and Palestinian economies.

*Public and Private Expenditures.* In Israel, current government expenditures are roughly 40 to 45 percent of GDP; over the past three decades, public policy has increasingly focused on welfare, social security, education, health, housing, and community services. In an April 2014 survey on the desired order of national priorities, the Israeli public gave top ranking to reducing socioeconomic gaps (47 percent) and second place to creating housing solutions at affordable prices (21 percent). These social sectors account for nearly 70 percent of total public expenditures; proportional and absolute increases have come at the expense of military expenditures and spending on economic services (e.g., direct subsidies, investments in transportation infrastructure).

Israel will face steadily increasing expenditures for social welfare as income disparities continue to grow and the relative sizes of the Haredi (strictly or ultra-Orthodox Jews characterized by rejection of modern secular culture) and the Palestinian citizens and residents of Israel increase. Expenditures for the settlements in the West Bank, including those around Jerusalem, account for more than 2 percent of government expenditures.

Government expenditures for the PA have grown rapidly over the past 20 years, nearly tripling between 1996 and 2012. More than 50 percent of the PA's spending is on defense and administration. Foreign aid continues to be essential for the PA to fund itself. A majority of the PA's self-generated revenue comes from tariffs on foreign imports and value-added taxes on Israeli goods and services, both of which Israel collects. Since the 2006 Palestinian legislative election, Israel has episodically withheld tax revenues for political reasons, most recently in response to the PA's move in January 2015 to join the International Criminal Court.

### Security

Israel will continue to face terrorist threats from Palestinian rejectionists, including Hamas; nonstate actors, such as al Qaeda and the Islamic State of Iraq and Syria (ISIS); and other state-supported external forces, such as Hezbollah. Other security concerns include threats to the stability of Israel's Arab neighbors and Iran's potential nuclear and long-range missile capabilities. Although Israel's security threats may increase over the decade, we assume, based on our interviews with security experts and our literature review, that the security environment will fall short of the sustained violent resistance experienced during the Second Intifada between 2000 and 2005. As a result, baseline expenditures for the Israel Defense Forces (IDF) will not change markedly in real terms.

Internally, the PA faces challenges from political and ideological rivals, such as Hamas. Additional threats include violence and vandalism from settlers and military incursions by Israel (e.g., combat operations in Gaza in 2014), both of which have been very costly over the years in terms of lives lost and infrastructure destroyed.

The burden on the Palestinian security apparatus is likely to grow, particularly in a two-state solution; requirements for training, equipment, and infrastructure will expand significantly. But the resources likely to be available to meet these requirements are not guaranteed. Given continued stalemate on the negotiating front, coupled with the measures Israel takes to maintain security—raids, an obvious military presence, checkpoints—episodic violent clashes with Gaza (e.g., in 2008, 2012, and 2014) seem likely to continue.

### Sociopsychological Dimensions of the Impasse

Our extensive literature review identified instances in which specific groups have suffered diagnosable mental health consequences of the long, uncertain, and occasionally violent Israeli-Palestinian conflict. In general, the degree of exposure and duration of exposure to violence were key factors driving the incidence of both mental health disorders and violence-related trauma symptoms. The literature has not yet accounted for the possible psychological effects of the 2014 war in Gaza.

## How Dimensions of Present Trends Would Change in Each Scenario

In our analysis, we consider how the key dimensions of present trends described above would change under the circumstances captured in our five scenarios.

### Economic Dimensions

We considered both direct costs (specific budgetary or financial expenditures related to the conflict) and opportunity costs (lost opportunities for fruitful activity resulting from the conflict).

For Israel, the primary direct costs stemming from the conflict include budgetary expenditures on settlements and security—e.g., military mobilizations. The largest opportunity cost is the impact of perceived instability in Israel on its investment and economic activity. Additional opportunity costs include lost opportunities for trade with the Palestinians and with the Arab world, reduced tourism, and less access to relatively affordable Palestinian labor for work in Israel.

For the Palestinians, including the West Bank, Gaza, and East Jerusalem, direct costs include the destruction of property, security direct costs, reductions in Palestinian labor working in Israel, restrictions on freedom of movement of goods and labor, banking regulations, and stipends for the families of prisoners held in Israel. The opportunity costs are more wide-ranging, mainly stemming from constraints caused by barriers to mobility, trade, and other economic activity, as well as lost economic activity from areas the Palestinians do not control.

### Security Dimensions

Israel's security posture includes a range of proactive and reactive mechanisms designed to deter security risks efficiently. Key elements include strategic warning (secured by the IDF in the West Bank and early warning stations in the West Bank and the Negev), tactical warning (including constant border patrols and periodic ground incursions), and a repertoire of proactive tactics designed to degrade potential terrorist infrastructure. Geographical separation between the populated areas of Israel and the Arab states to the east provides a buffer zone. This, coupled with strategic warning, provides strategic depth. Israel maintains the ability to respond rapidly to imminent threats, to preposition forces for deterrence, and to restrict freedom of movement in order to disrupt would-be terrorist activity within the West Bank. Israel uses a combination of such mechanisms to maintain security on the West Bank at relatively low cost to Israel.

In Palestinian security planning, strategic warning, a buffer zone, and strategic depth do not play a role; instead, funding, the requirements for a justice system, and the importance of chain of command are central. Developing responsible internal and external security structures, building basic capabilities, acquiring the essential tools of security, and negotiating authorities within the constrained Israeli framework are among the major challenges to effective Palestinian security.

On the security dimension, the imbalance in power between Israel and the Palestinians is enormous—Israel controls virtually all aspects of security because it believes that Palestinian security forces lack the capacity to address terrorism and did not do so in Gaza. Thus, Israel supports the Palestinian security services or constrains them as it sees fit.

## Assumptions for Five Alternative Scenarios

For each scenario, we make a series of economic and security assumptions. Tables S.2 and S.3 describe the economic assumptions for, respectively, the Israelis and the Palestinians. Table S.4 summarizes our security assumptions for each scenario.

**Table S.2**
**Economic Assumptions for Israelis**

| | Scenario | Two-State Solution | Unilateral Withdrawal | | Nonviolent Resistance | Violent Uprising |
|---|---|---|---|---|---|---|
| | | | Coordinated | Uncoordinated | | |
| Direct costs | 1. Security | No change | No change | Defense expenditures increase by 1% | No change | Defense expenditures increase by 3% |
| Direct costs | 2. Settlement | 100,000 settlers leave West Bank with proportional (16%) reduction in annual costs; relocation costs paid for by international community | 60,000 settlers leave West Bank with proportional (10%) reduction in annual costs; 75% of relocation costs paid for by international community | 30,000 settlers leave West Bank with proportional (5%) reduction in annual costs; 0% of relocation costs paid for by international community | No change | No change |
| Direct costs | 3. Palestinian services | No change | No change | No change | No change | Israel pays for Palestinian health, education, and social welfare |
| Opportunity costs | 1. Instability and uncertainty | 15% increase in investment and labor productivity for 4 years | No change | 5% decrease in investment rate for 4 years | 10% decrease in investment rate for 4 years | 20% decrease in investment, 100% reduction in total factor productivity, 50% reduction in labor market growth for 4 years |
| Opportunity costs | 2. Boycott, divestment, and sanctions (BDS) | No change | No change | No change | 2% reduction in GDP | No change |
| Opportunity costs | 3. Tourism | 20% increase | 5% increase | 5% decrease | 10% decrease | 25% decrease |
| Opportunity costs | 4. Arab world trade | Trade with the greater Middle East triples | No change | No change | No change | No change |
| Opportunity costs | 5. Palestinian trade | 150% increase | 10% increase | No change | No change | 15% decrease |
| Opportunity costs | 6. Palestinian labor in Israel | Permits increase by 60,000 | Permits reduced by 30,000 | Permits reduced by 30,000 | Permits reduced by 30,000 | Palestinian labor in Israel stopped |

**Table S.3**
**Economic Assumptions for Palestinians**

| | Scenario | Two-State Solution | Unilateral Withdrawal | | Nonviolent Resistance | Violent Uprising |
|---|---|---|---|---|---|---|
| | | | Coordinated | Uncoordinated | | |
| Direct costs | 1. Destruction of property | No change | No change | No change | No change | $1.5 billion in damage to capital stock |
| | 2. Territorial waters | Access for resource extraction | No change | No change | No change | No change |
| | 3. Palestinian labor in Israel | Permits increase by 60,000 | Permits reduced by 30,000 | Permits reduced by 30,000 | Permits reduced by 30,000 | No Palestinian labor in Israel |
| | 4. Freedom of movement | All costs removed | All costs removed | No change | 25% increase | 100% increase |
| | 5. Access to social services | 25% reduction in costs | No change | No change | 25% increase in costs | 50% increase in costs |
| | 6. Banking regulations | 50% reduction in costs | No change | No change | No change | 50% increase in costs |
| | 7. Prisoners in Israel | All political prisoners released | No change | No change | 10% increase in number of prisoners | 100% increase in number of prisoners |
| Opportunity costs | 1. Control of territory | Full control of land vacated by IDF and 100,000 settlers | Full control of land vacated by IDF and 60,000 settlers | Full control of land vacated by 30,000 settlers | No change | No change |
| | 2. Access to water | Unlimited access at market price | Unlimited access at market price | No change | No change | No change |
| | 3. Barriers to trade | 50% reduction in transaction costs | 10% reduction in transaction costs | No change | 25% increase in transaction costs | 50% increase in transaction costs |
| | 4. Licensing | Elimination of licensing restrictions | No change | No change | No change | No change |
| | 5. Tourism and travel | Visa restrictions lifted | No change | No change | No change | No change |
| | 6. Dissolution of PA | No change | No change | No change | No change | PA collapses |
| | 7. Investment in public and private infrastructure | Sufficient for all new economic opportunities | Sufficient for all new economic opportunities | Sufficient for 50% of new economic opportunities | No change | No change |

**Table S.4**
**Security Assumptions**

| | Scenario | Two-State Solution | Unilateral Withdrawal: Coordinated | Unilateral Withdrawal: Uncoordinated | Nonviolent Resistance | Violent Uprising |
|---|---|---|---|---|---|---|
| Israeli | 1. Strategic warning | No change | No change | No change | No change | No change |
| | 2. Tactical warning | Reduced | Reduced | Reduced | No change | Increased |
| | 3. Buffer zone | Reduced | No change | No change | No change | No change |
| | 4. Strategic depth | Reduced | No change | No change | No change | Reduced |
| | 5. Mobility | Reduced | Reduced | No change | No change | Increased |
| | 6. Border security | No change | No change | No change | No change | Increased |
| | 7. Liaison | Increased | No change | No change | Decreased | Reduced |
| Palestinian | 1. Force size | Significant expansion in new mission areas | Increased | Need increased | No change | Significant decrease and/or destruction |
| | 2. Force structure | Significant growth | Increased | Need increased | No change | Reduced |
| | 3. Funding | Increased | Increased | No change | No change | Reduced |
| | 4. Chain of command | Increased | No change | No change | No change | Dispersal of authority; local militias dominate |
| | 5. Liaison relationships | Increased | No change | No change | Reduced | Cut off |
| | 6. Freedom of movement | Increased | Increased | No change | No change | Decreased |
| | 7. Border security | Increased | No change | No change | No change | Decreased |
| | 8. Justice | Increased | Increased | Somewhat increased | No change | No change |

For example (from Table S.2), in a two-state solution, Israel would see a 16-percent reduction in the direct costs of the settlements and would benefit from a 15-percent increase in investment because of perceived increased stability in the region. In contrast, in a violent uprising, Israel would see a 3-percent increase in average defense expenditures in each of the ten years (a direct cost) and a 25-percent decrease in tourism (an opportunity cost).

Analogously (from Table S.3), in a two-state solution, the Palestinians would benefit from full control of land and sea, except the Jordan River Valley, and a 50-percent reduction in transaction costs due to the removal of trade barriers. In the case of a violent uprising, Palestinians could suffer up to $1.5 billion in direct costs as a result of damage to capital stock and a 100-percent increase in transaction costs due to very stringent restrictions on movement.

## Economic Costs and Benefits for Each Scenario

The aggregate economic costs and benefits of each scenario in 2024, compared with outcomes of present trends, are shown in Figures S.2 and S.3 and Tables S.6 and S.7 (presented at the end of this summary). All results compare outcomes in 2024 relative to what the outcomes would have been if present trends had continued. The aggregate economic changes in GDP are reported in Figures S.2 and S.3, which combine the

**Figure S.2**
**Change in Economic Costs in 2024 for the Five Scenarios Studied Relative to Present Trends as a Percentage of GDP**

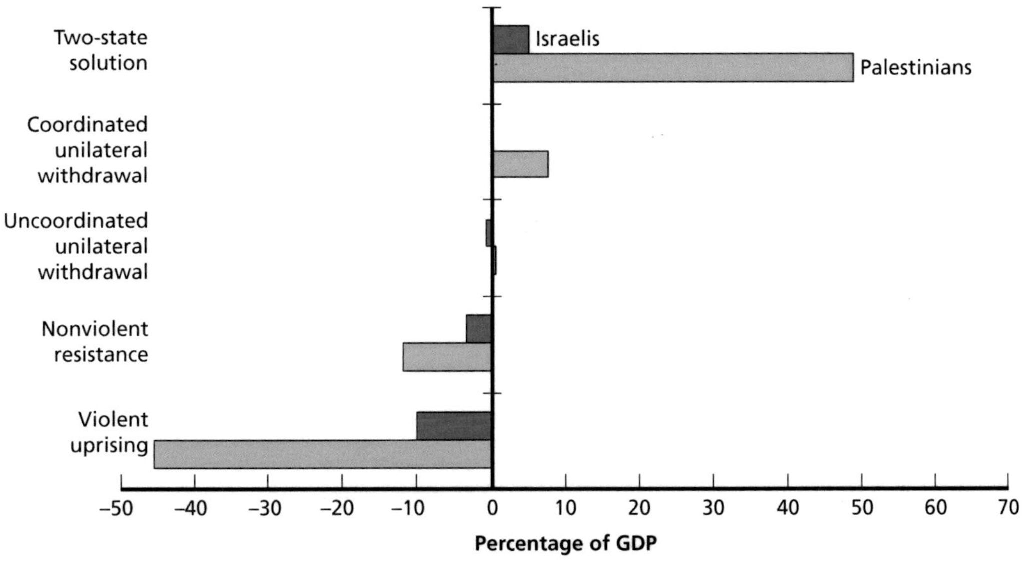


RAND *RR740/1-1-S.2*

**Figure S.3**
**Change in Economic Costs in 2024 for the Five Scenarios Studied Relative to Present Trends in GDP Per Capita (in U.S. dollars)**

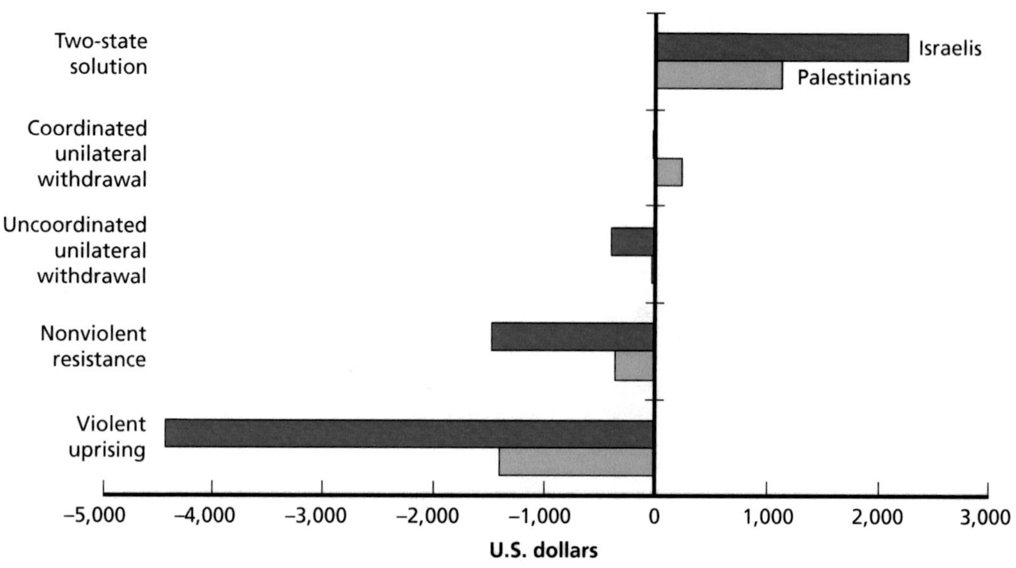


RAND *RR740/1-1-S.3*

direct and opportunity cost effects on GDP based on a conservative assumption about the fiscal multiplier.

The **two-state solution** assumes that the Israelis and Palestinians reach a final status accord agreement that is generally based on the Clinton Parameters. Israelis will withdraw to the 1967 borders except for mutually agreed-upon swaps, and we assume that 100,000 settlers will be relocated from the West Bank to Israel. The Palestinians will gain full control of Areas B and C and the ability to exploit the mineral resources there (see Map 1). All trade and travel restrictions on the Palestinians will be lifted, and perhaps as many as 600,000 refugees might return to the WBG in a phased manner. Israeli settlers withdraw from the West Bank except for the agreed-upon swap areas, and the international community pays most of the costs for relocating settlers. Israel's security is guaranteed by the international community, and investment in both Israel and Palestine is forthcoming to take advantage of a new, stable climate and the opportunities that peace brings. Arab country sanctions on Israeli trade are lifted, and Israeli trade with Arab countries increases rapidly.

A two-state solution produces by far the best economic outcomes for both Israelis and Palestinians. However, Israel would benefit more than the WBG in absolute terms. Our analysis suggests that, as a result of the effect of changes in direct and opportunity costs on GDP in the year 2024, Israel's GDP would increase by $23 billion over what it would have been if present trends had continued, while GDP in the WBG would be $9.7 billion larger. The average Israeli would see his or her income in 2024 increased by about $2,200 (about 5 percent), while the average Palestinian's income would rise by about $1,100 (about 36 percent).

**Coordinated unilateral withdrawal** assumes that Israel coordinates withdrawal from much of the West Bank with the Palestinians, who cooperate, and with the international community; withdrawal is implemented in stages over ten years. We assume that 60,000 settlers would be withdrawn and that the lands they occupied would come under full Palestinian control and be available for full economic use. Seventy-five percent of the cost of settler evacuation would be paid for by the international community and 25 percent by Israel. West Bank checkpoints and other barriers to trade would be greatly reduced, and other transaction costs to international trade would fall by 10 percent. We assume that investment needed to exploit the new economic opportunities would be forthcoming from a combination of the diaspora, international direct investment, and/or donor aid.

Israel's security footprint and costs would change little, while Palestinian security costs would increase significantly to cover expanded responsibilities. Labor permits for Palestinians to work in Israel would be decreased by 30,000 as Israel seeks to disconnect its economy from that of the Palestinians.

Israel experiences little if any economic effect because the various positive and negative factors cancel each other; Palestinians see $1.5 billion growth in GDP by the tenth year, with per capita GDP about 8 percent larger than in present trends, reflecting the economic potential opened up in Area C and reduction in internal and external barriers to economic activity and trade, offset somewhat by a decrease in income as a result of decreased employment in Israel. Direct costs for both parties are relatively small.

The **uncoordinated unilateral withdrawal** scenario is consistent with the belief of many that neither the Palestinians nor the international community is likely to agree to a policy that does not address any of the Palestinians' long-standing aspirations. In this scenario, Israel nevertheless proceeds on its own to cede control of some of the West Bank, but, in this case, only 30,000 settlers will willingly leave areas on the West Bank. In addition, Israel will have to pay 100 percent of their relocation costs—nearly $850 million per annum in total resettlement costs. Although there will be a reduction in some direct Israeli security costs related to the settlements, we expect little overall change in security costs, as the IDF will have more or less the same responsibilities, and there could be increased unrest from both settlers and Palestinians over the ten years examined.

Overall, Israeli GDP, compared to present trends, will fall by about $4 billion in 2024, only a 0.9-percent change in GDP per capita. Palestinians will similarly see a slight 0.5-percent reduction in GDP per capita, with the negative impact of lost Palestinian labor in Israel overshadowing the benefit of a slight increase in economic opportunities in Area C.

The **nonviolent resistance** scenario assumes that the Palestinians take actions to put economic and international pressure on the Israelis. This includes efforts in the UN and the International Criminal Court and boycotts of Israeli products in the WBG. We also assume growth of the BDS movement around the world, but primarily in Europe. We assume that Israel will respond with a number of measures, including reducing the number of work permits issued by Israel to Palestinians to work in Israel by 30,000, increasing internal and external barriers to trade and movement, and periodically withholding payment of taxes that it collects for the Palestinians. In this scenario, security costs for both Palestinians and Israelis are likely to rise. Some feel that the present trends baseline we have defined has already evolved in part into nonviolent resistance.

As a result of increased opportunity costs, nonviolent resistance will cause Israel's GDP to fall by $15 billion below present trends, a reduction of 3.4 percent per capita (about $1,500). Palestinians experience a reduction in GDP of $2.4 billion, or 12 percent per capita (about $370). The drop in Israel's GDP results primarily from reduced international investment and tourism because of perceived instability in the region and from a broader BDS movement in Europe. Palestinians suffer as a result of Israeli destruction of property and retaliation, which increases barriers and transaction costs to trade, and from decreased income because fewer Palestinian workers are in Israel.

In a **violent uprising** scenario, violence erupts—perhaps beginning in Gaza but spreading to the West Bank and possibly involving such foreign actors as Hezbollah in the north. We do not model this scenario as a repeat of the Second Intifada, but how it starts and its ultimate form are hard to project. Israelis would respond with actions designed both to punish the Palestinians and to establish tighter control.

We assume that these actions would cause the PA to collapse; Israel would then have to assume responsibility for essential functions that the PA currently provides, such as security in Area A, and shoulder the costs for health, education, and social services. This is the only scenario in which we modeled the collapse of the PA, but Israeli actions, such as withholding tax receipts and/or the withering of international support, could cause collapse of the PA in any of the other scenarios except a two-state solution.

A return to violence would have strong negative effects on both parties as a result of opportunities lost. As a result of increases in both direct costs and opportunity costs, GDP would fall $9.1 billion for the Palestinians and an estimated $45 billion for the Israelis, as compared to present trends. GDP per capita would fall by 46 percent in the WBG and by 10 percent in Israel. Israel's drop in GDP stems from effects of increased security costs and the effects of an unstable environment on investment and tourism.

Palestinians suffer because of the reduction in trade and economic activity as Israel increases barriers to both, the collapse of the PA, the destruction of homes and infrastructure, and the elimination of Palestinian labor in Israel. Much of the costs resulting from the dissolution of the PA will be in the areas of security, health, and education; Israel will likely have to bear many of these costs, amounting to $2.4 billion per annum as of 2024.

The relative importance of different opportunity and direct costs for the two economies is highlighted in Table S.5, which reports the economic differences between the present trends and each of the five scenarios as of 2024 (i.e., the scenario minus present trends). The economic differences are disaggregated, with the effects of each type of opportunity or direct cost on GDP reported separately. *In each case, a positive number indicates that the economy is wealthier along that dimension for that scenario than in present trends.* In the table, the aggregate changes in GDP because of changes in opportunity and direct costs are broken out separately.

For Palestinians, direct costs include a wider range of restrictions, but they typically reflect external restrictions that the Israelis can place on the Palestinian economy. The largest costs across all the scenarios are the restrictions that Israel can place on the flow of labor, though the destruction of property is expected to be quite large in the violent uprising scenario. The effect of opportunity cost changes on GDP are also larger for the Palestinians; these costs are dominated by restrictions on trade, though the difference between opportunity costs and direct costs is smaller.

The effects on GDP of the changes in opportunity costs far outweigh the importance of direct costs in almost all scenarios, especially with respect to GDP per capita, except in the two unilateral withdrawal scenarios.

### Ten-Year Aggregation Across Scenarios

The difference between the economies after ten years in each of the five scenarios captures the difference in only the final year of our ten-year counterfactual analysis. However, these economies will be either richer or poorer across each of the ten years for each of the scenarios. Therefore, we also calculate the aggregate ten-year difference in GDP under the assumption that the effects of changes in costs and benefits of each of the scenarios are realized gradually (Figure S.4).

The two-state solution results in combined ten-year benefits to Israel of $123 billion, or a little less than half of Israel's 2014 GDP; the total benefit for the Palestinians is $50 billion, more than three times the size of their 2014 GDP. The combined total of wasted economic opportunity for both parties is more than $170 billion. Mirroring the year 2024 results, the aggregate ten-year figures in coordinated and uncoordinated unilateral withdrawal are very small. Nonviolent resistance would cost Palestinians $12 billion over ten years and the Israelis $80 billion; a violent uprising would cost Israel $250 billion (slightly less than its 2014 GDP) and the Palestinians $46 billion (more than three times their 2014 GDP).

As was the case for outcomes in the year 2024, the ten-year total economic effects are much greater for the Palestinians than for Israel because the Israeli economy and per capita income are so much larger.

**Table S.5**
**Itemized Economic Differences Between Present Trends and the Five Scenarios in 2024 (changes in GDP in millions of U.S. dollars)**

| Israelis | | | | | | Palestinians | | | | | |
|---|---|---|---|---|---|---|---|---|---|---|---|
| | | Unilateral Withdrawal | | | | | | Unilateral Withdrawal | | | |
| | Two-State Solution | Coordinated | Uncoordinated | Nonviolent Resistance | Violent Uprising | | Two-State Solution | Coordinated | Uncoordinated | Nonviolent Resistance | Violent Uprising |
| All Costs | | | | | | | | | | | |
| **Total change in GDP from all costs** | **$22,800** | **–$180** | **–$4,010** | **–$15,000** | **–$45,100** | **Total change in GDP from all costs** | **$9,700** | **$1,510** | **–$110** | **–$2,380** | **–$9,080** |
| Direct Costs[a] | | | | | | | | | | | |
| Security | $0 | $0 | –$130 | $0 | –$340 | Destruction of property | $0 | $0 | $0 | $0 | –$750 |
| Settlement | $90 | –$170 | –$420 | $0 | $0 | Territorial waters | $280 | $0 | $0 | $0 | $0 |
| Palestinian services | $0 | $0 | $0 | $0 | –$830 | Palestinian labor in Israel | $360 | –$180 | –$180 | –$180 | –$760 |
| | | | | | | Freedom of movement | $80 | $80 | $0 | –$20 | –$80 |
| | | | | | | Access to services | $10 | $0 | $0 | –$10 | –$30 |
| | | | | | | Banking regulations | $4 | $0 | $0 | $0 | –$4 |
| | | | | | | Prisoners in Israel | $90 | $0 | $0 | –$20 | –$180 |
| **Total change in GDP from direct costs** | **$90** | **–$170** | **–$550** | **$0** | **–$1,170** | **Total change in GDP from direct costs** | **$820** | **–$110** | **–$180** | **–$230** | **–$1,800** |

**Table S.5—Continued**

| | Israelis | | | | | | Palestinians | | | | |
|---|---|---|---|---|---|---|---|---|---|---|---|
| | Two-State Solution | Unilateral Withdrawal | | Nonviolent Resistance | Violent Uprising | | Two-State Solution | Unilateral Withdrawal | | Nonviolent Resistance | Violent Uprising |
| | | Coordinated | Uncoordinated | | | | | Coordinated | Uncoordinated | | |
| Opportunity Costs | | | | | | | | | | | |
| Instability and uncertainty | $9,100 | $0 | –$2,530 | –$5,050 | –$39,200 | Control of territory | $480 | $290 | $70 | $0 | $0 |
| BDS | $0 | $0 | $0 | –$8,780 | $0 | Access to water | $780 | $470 | $0 | $0 | $0 |
| Tourism | $1,120 | $280 | –$280 | –$560 | –$1,400 | Barriers to trade | $4,300 | $860 | $0 | –$2,150 | –$4,300 |
| Arab world trade | $5,580 | $0 | $0 | $0 | $0 | Licensing | $30 | $0 | $0 | $0 | $0 |
| Palestinian trade | $5,580 | $370 | $0 | $0 | –$560 | Tourism and travel | $770 | $0 | $0 | $0 | $0 |
| Palestinian labor in Israel | $1,320 | –$660 | –$660 | –$660 | –$2,790 | Dissolution of PA | $0 | $0 | $0 | $0 | –$2,980 |
| **Total change in GDP from opportunity costs** | **$22,700** | **–$8** | **–$3,460** | **–$15,000** | **–$43,900** | **Total change in GDP from opportunity costs** | **$6,360** | **$1,610** | **$70** | **–$2,150** | **–$7,280** |
| | | | | | | **Other costs: Return of refugees** | **$2,510** | **$0** | **$0** | **$0** | **$0** |

NOTE: Data may not match total because of rounding.

[a] In order to translate changes in direct costs into changes in GDP, we assume (1) a fiscal multiplier of 0.5 and (2) that any changes in direct costs are reflected by a 1:1 change in government expenditures.

**Figure S.4**
**Ten-Year Total Combined Change in GDP from 2014 Through 2024 for the Five Scenarios Studied (in billions of U.S. dollars)**

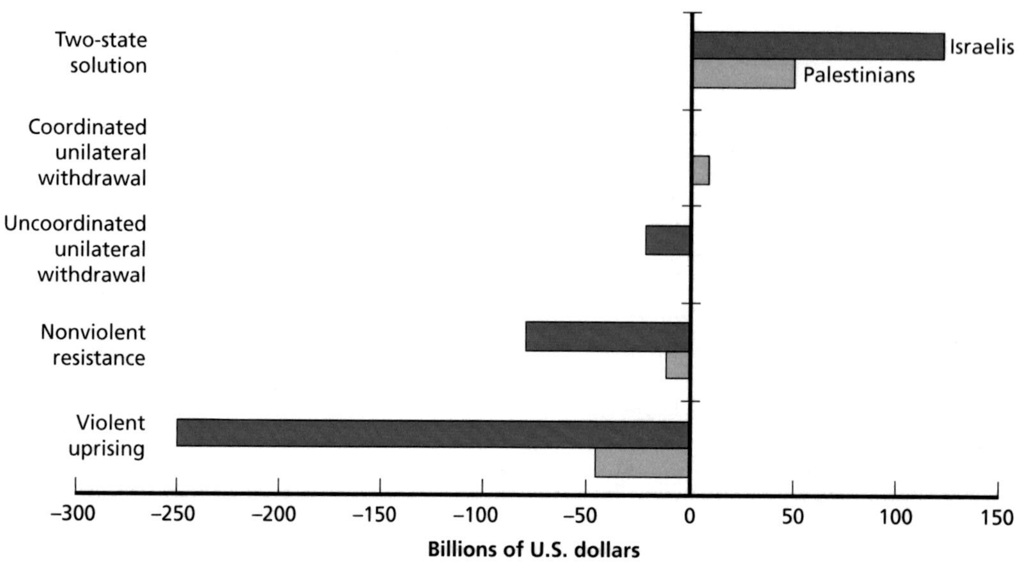


RAND *RR740/1-1-S.4*

In a two-state solution, Israel would benefit from increased direct investment in the domestic economy and from new trading opportunities with the Arab world. Israel's only short-run direct benefit from peace results from slightly reduced payments to support the settlements, which we assumed here would be removed from the West Bank with substantial international support; however, many settlement blocs would become part of Israel with agreed-upon border adjustments as per the Clinton Parameters, noted earlier. Palestinian benefits or costs (depending on the scenario) stem from the impacts on trade opportunities, reflected as a reduction or increase in the costs of producing and moving inputs and goods. The biggest direct effect for the WBG in any scenario stems from impact on employment opportunity in Israel. All of the economic results are dependent on and follow directly from our analytic assumptions.

Tables S.6 and S.7, presented at the end of this summary, list the costs and benefits of present trends compared with each of the five scenarios over the period 2014–2024.

## Security Outcomes

We examine and calculate direct security costs for Israel and for the Palestinians, as well as their security needs and frameworks. We also examine but do not quantify the increase or reduction in perceived security risk resulting from each scenario. We draw the following conclusions:

- *In the short term, security costs are unlikely to fall significantly for either party under any scenario.* But, under a two-state solution, Palestinian security expenditures would likely rise rapidly as the role of the Palestinian National Security Forces expands substantially both internally and externally. In uncoordinated withdrawal, nonviolent resistance, or a violent uprising, we expect both Israeli and Palestinian security costs to rise.
- *Israel sees unchanged/increased security risks under any scenario.* Any deviation from its current approach to security involves increased uncertainty and greater perceived security risk.

## Costs and Investments for the International Community

Each of the five scenarios also has cost implications for the international community—in particular, for the United States and Europe, which have provided financial and political support to both Israel and Palestinians since World War II. The net ten-year cost of each scenario is summarized in Tables S.6 and S.7. Given our counterfactual approach, we report only net changes from the status quo—that is, we assume that large military aid flows to Israel and humanitarian aid to the Palestinians will continue at present trends rates.

### Support to Israel

Israel, the largest recipient of official U.S. foreign aid since World War II, has received about $118 billion to date—recently, about $2.6 billion per year. Aid includes defense assistance and a variety of nondefense support, including grants and emergency assistance during economic slowdowns and other geopolitical events. Israel benefits from U.S. budget appropriations related to many other defense programs and various technology transfer programs. Charitable donations from private U.S. organizations and individuals are also a major source of financial support to Israeli institutions; donations are usually made through U.S. tax-exempt organizations. We assume that aid flows to Israel primarily from the United States will continue at current levels.

Funding from the international community to Israel in a two-state solution scenario is primarily to help pay for relocating settlers who move out of the West Bank and the increased security costs that the international community, and the United States in particular, assume as security guarantees in any final settlement. Because the exact nature of those guarantees is unknown, we do not cost them here. In the two-state solution, we assume that funding would total an estimated $30 billion across the ten years; funding would total $13.5 billion in the coordinated unilateral withdrawal scenario.

### Support to Palestine

The historical international contribution to the Palestinians has been predominantly development aid and direct budget support. Until the Oslo Accords, U.S. support to Palestine flowed primarily through the UN Relief and Works Agency for Palestine Refugees in the Near East ($1.6 billion between 1950 and 1991). After the Oslo Accords, foreign aid increased dramatically, with expanded support for the PA, which assumed responsibility for many of the social services formerly administered by nongovernmental organizations. After the Second Intifada, the Palestinian economy became very reliant on the direct support of the international community, which has continued.

In the two-state solution scenario, we assume that aid and direct support will continue at similar levels, and we focus on the types of new resources that the Palestinians would require from the international community to take advantage of economic opportunities that emerge in each scenario. Such resources would include a mixture of private and public investment to exploit new economic opportunities in Area C and new trade opportunities. In the two-state solution scenario, we also assume that the international community would provide assistance to fund the new investments that will be needed to repatriate refugees returning to the new Palestinian state from abroad.

In a violent uprising scenario, we estimate that the international community may actually have reduced expenditures. In this case, the flow of aid and direct support from the international community would slow dramatically as Israel assumes the costs for health, education, and social affairs for the Palestinians following the PA's collapse.

Israel will also need significant investment from international and domestic sources to take full advantage of the economic opportunities that peace would bring. We do not try to calculate the amount that would be international in origin: Israel's well-developed capital markets will enable these funds to flow smoothly at market-clearing prices. These funds, the amounts of which are likely to be quite large, are not included in our calculations.

## Other Important Noneconomic Factors

Multiple studies (including our own) have demonstrated that the two-state solution is clearly the best solution for both parties, and violence the worst. So why has the Israeli-Palestinian impasse endured? Either the parties do not properly recognize the economic benefits of an agreement, or the economic benefits of an agreement have not been and may not be high enough to outweigh the imputed costs of other factors associated with the present trends, including the perceived costs of such intangible factors as distrust and fear of relinquishing some degree of security.

Based on our literature reviews and interviews, we suggest some of the factors, many interrelated, that may constitute barriers to resolving the impasse. We do not try to assess the relative importance of any one of these factors.

***Power imbalance:*** *Israel, the country with by far the greater power, has a smaller economic incentive to diverge from the present trends. The percentage changes in income for the average Israeli are far smaller than changes for the average Palestinian.*

A singular feature of the Israeli-Palestinian relationship is the power imbalance between the two parties: Israel dominates the region both militarily and economically. This imbalance is true in every scenario.

***Economic incentives:*** *Israelis have less of an economic incentive than the Palestinians do to resolve the impasse.*

Since the Israeli economy and the average per capita income are more than 15 times larger than is the case in the Palestinian economy, absolute changes have a relatively small effect on individual Israelis. The opposite is true for the Palestinian economy. In the case of security, Israel is by far the dominant force in all respects. As noted earlier, because its economy is so much larger than that of the Palestinians, Israel has larger absolute gains from peace or losses from violence, but, in percentage terms, the effect on the average Israeli is much less.

***Security management:*** *Israel has learned how to manage security vis-à-vis the Palestinians at relatively low cost. Diverging from present trends entails significant uncertainty that influences both parties as they consider final status accord issues.*

The scenario of most interest from a security perspective is a two-state solution. For understandable historical reasons, Israel is very risk averse when it comes to security. Any change from the status quo brings uncertainty and, therefore, perceived increased security risk.

The core tenet of Israel's security doctrine is that it must rely only on itself to ensure the state's survival, and not on the United States, although American assistance and relationships with the United States certainly contribute significantly to Israel's security. Israel does not believe that the PNSF can maintain security to Israeli standards without an IDF presence, and Israeli negotiators reportedly have asserted that Israel will not accept international forces as a compensating element. Many in Israel insist on retaining a security corridor and control of the Jordan River and other border crossings. Israel's lack of trust in the Palestinians and its doubt about the international community's commitment to its security appear to outweigh any potential economic benefits that could flow from taking an alternative trajectory.

The Palestinians would presumably be eager to assume new responsibilities in a two-state solution, but Western experts believe that they lack the experience, manpower, and resources to fulfill all of them quickly. Internally, Palestinian security forces would face sharply increased responsibilities, assuming functions currently handled by the IDF, as well as maintaining peace in Gaza. The degree of cooperation from Hamas and other non-Fatah parties will play a critical role in determining the ultimate effectiveness of the security force and arrangements.

***Lack of political consensus:*** *Deep political and religious divisions make it more difficult for either Palestinians or Israelis and their leaders to garner popular support for accepting the compromises required to break the impasse. Subgroups in each population are powerful enough to make change difficult.*

Both Israelis and Palestinians living in the WBG are deeply divided politically and religiously. Fatah and Hamas have significant levels of support in both areas. They also have profoundly different attitudes about Israel and divergent approaches to resolving the conflict, making cooperation between them very difficult. Israel also has many deep and complex political divisions, in part stemming from religious identity and the dominant political perspective of each group.

***Lack of leadership:*** *Neither side believes that it has a partner with which to negotiate peace, and neither side appears to have the leadership necessary to create a new vision and transform it into reality.*

Prime Minister Benjamin Netanyahu has often said that Israel has no partner for peace, a situation many Israelis do not see changing as long as there is a coalition government that includes Hamas. The Palestinians also feel that they lack a partner, and, after Netanyahu's March 16, 2015, declaration that there would be no Palestinian state while he was prime minister, Palestinians have little expectation that Israel under Netanyahu will negotiate a final status accord agreement.

***Regional instability:*** *The Middle East is plagued with upheaval and instability, and there is little on the region's ten-year horizon likely to change this situation.*

The region around Israel in the Middle East has been in chaos for some time, and, for a number of reasons, the political terrain is shifting. The centrality of the Israeli-Palestinian conflict today is receding, overshadowed by the rise of ISIS, the metastasizing Syrian civil war, the collapse of any semblance of governance in Libya and Yemen, Shi'a-Sunni tensions, Egyptian and Gulf State hostility to the Muslim Brotherhood, and Iran's nuclear and regional ambitions. However, the Israeli-Palestinian conflict retains significant sway in Arab public opinion, and periodic outbursts of violence will likely continue, intermittently propelling the issues embodied in the conflict to the forefront of international concern.

***Conflicting narratives:*** *The historical narratives of Israelis and Palestinians, although parallel, are in fundamental conflict with each other. The clash of these narratives, including the increasingly important roles of religion and ideology, significantly enhances the probability that the impasse will continue.*

Fear and mistrust have led Israel to approach the peace process with great caution. Israelis do not trust the Europeans or the international community to stand behind them. They are very reluctant to trust Palestinians in particular and other Arab states more generally. Some Israelis also believe that Israel's destiny is to incorporate all of ancient Judea and Samaria into their state. These feelings make accommodation with the Palestinians exceedingly difficult.

Palestinians seek fulfillment of their long-standing aspirations to national independence and sovereignty. They view East Jerusalem as intrinsically Palestinian territory and Israeli settlements elsewhere in the West Bank as expropriation in violation of established international law. As among Israelis, an important segment of Palestinians frames their national struggle in religious, as well as nationalist, terms. Most Palestinians view Israeli actions as collective punishment or humiliation and subjugation intended to penalize and suppress their national aspirations. Palestinians are deeply pessimistic about whether Israel is negotiating in good faith, noting that the scope and scale of Israeli settlement beyond the Green Line expanded faster after the Oslo Accords than before (see Map 2). Nurturing an environment of mistrust is the fact that Palestinians and Israelis now have little or no direct contact with each other except in the specific context of conflict.

***International donor enabling:*** *The cost of the status quo to both Israelis and Palestinians would be significantly higher were it not for donor aid that has, to some extent, insulated both parties from the total cost of the impasse and lessened incentives to seriously pursue a final status accord agreement.*

Israel continues to receive the world's largest share of official U.S. foreign assistance funding. International aid coming chiefly from the United States, the European Union, and the UN primarily through the UN Relief and Works Agency for Palestine Refugees in the Near East has been a critical component of support for the PA since its formation in the 1993 Oslo Accords.

## Where Do We Go from Here?

Key characteristics of Israeli, Palestinian, and international policymaking, strategic thinking, political dynamics, demographics, and social dynamics will shape Israeli and Palestinian relations in the coming years. These trends will be part of a sustained feedback loop, influencing the responses of all parties in ways that reinforce these characteristics. The trends will, in turn, reinforce, perpetuate, and intensify the cycle of action and reaction.

Each cycle will progressively close off options that the parties might have had to break the cycle in potentially favorable ways. As options fade, parties become trapped in circumstances that may be quite far from the outcome they had imagined at earlier stages.

Outcomes projected in our scenarios may already be appearing. The cycle of tit-for-tat moves to pressure the other side has begun in earnest. The PA's UN and International Criminal Court bids were among the assumptions of our nonviolent resistance scenario. In response, Israel withheld Palestinian taxes for four months and has been considering other actions that raise economic costs to the Palestinians. For their part, Palestinians may respond by boycotting Israeli goods and services collectively (this is already a trend, but, in response to Israeli collective action, it may become an official policy). The continuing impasse will also likely enhance the BDS movement and result in a greater economic effect on Israel. The Israelis may, in turn, respond with boycotting Palestinian goods and services. The United States is reassessing its aid to the PA, and some members of Congress have introduced legislation to cut off aid completely.

Six broad trends will exert powerful influences on this cycle of action and reaction:

- *Continued settlement expansion* will make it increasingly difficult and costly to move settlers and resolve the impasse.
- An *open media environment* allows instant communication and worldwide exposure to events as they unfold, making it difficult for parties to disseminate their own interpretations of events.
- *The technology of war and terrorism* will continue to evolve rapidly. With external state actor support, range and guidance systems for terrorist rockets will improve and will be pitted against Israeli improvements in antimissile technology and its Iron Dome system.
- *Public opinion may be shifting.* Young American Jews feel less affinity for Israel and its policies than previous generations and are more apt to criticize its policies. Among British and French publics, strong pluralities report sympathizing more with Palestinians than Israelis. European parliaments, including those in the United Kingdom, Sweden, Ireland, and France, are voting to support recognizing a Palestinian state. The BDS movement has not yet had a significant negative effect on Israel. However, the movement is growing, particularly in Europe, Isra-

el's largest trading partner, and some Israeli leaders have warned that the movement's effects could have substantial detrimental effects on the economic welfare of Israelis.

- *Regional instability* continues unabated, including the rise of ISIS; civil wars in Syria, Libya, and Yemen; Iran's regional ambitions; and the collapse of governance in Libya and the Sinai.
- *Demographic trends* foreshadow a Palestinian majority in the territory comprising Israel and the WBG—either today or in the near future—and a Palestinian majority in 30 years, even if the population of Gaza is not included. Then, it has been suggested, Israel would face a core policy choice: whether to be a Jewish state with a predominantly Jewish population living side by side with a Palestinian state, a democratic state with a diverse citizenry that is treated equally, or a state without a Jewish majority that comprises the lands known as Israel and all the land between the Jordan River and the Mediterranean Sea.

Some scholars say that, given recent trends, the two sides are marching toward a one-state solution unless Israel opts for unilateral withdrawal, an alternative that also becomes increasingly problematic as West Bank settlements expand. Exactly what the one-state options are and how a single state—even a federation—would operate has not been extensively examined. Research is needed on how—and whether—a one-state solution could be structured in a way that preserves democratic principles.

A potential diversion from the current trajectory could come about if the parties were to radically change the way they currently view the impasse. But, to achieve that, dramatic policy intervention by all would be needed.

We hope our work can help Israelis, Palestinians, and the international community understand more clearly how present trends are evolving and recognize the costs and benefits of alternatives to the current destructive cycle of action, reaction, and inaction.

**Table S.6**
**Summary of the Economic Change of Present Trends for Israel in the Five Scenarios in 2024**

| | | | | 2024 Scenarios | | | | |
|---|---|---|---|---|---|---|---|---|
| | | | | | Unilateral Withdrawal | | | |
| | | 2014 | 2024 Extrapolation | Two-State Solution | Coordinated | Uncoordinated | Nonviolent Resistance | Violent Uprising |
| Opportunity costs only | Population (millions)[a] | 8.2 | 10.2 | 10.2 | 10.2 | 10.2 | 10.2 | 10.2 |
| | Total GDP (U.S.$, billions) | $295 | $439 | $462 | $439 | $436 | $424 | $395 |
| | Change in GDP (U.S.$, billions) | – | – | $22.7 | $0.0 | –$3.5 | –$15.0 | –$43.9 |
| | Change in GDP (%) | – | – | 5.2% | 0.0% | –0.8% | –3.4% | –10.0% |
| | GDP growth rate (average) | – | 4.1% | 4.6% | 4.1% | 4.0% | 3.70% | 3.0% |
| | GDP per capita (U.S.$) | $35,900 | $43,300 | $45,500 | $43,300 | $42,900 | $41,800 | $39,000 |
| | Change in GDP per capita (U.S.$) | – | – | $2,240 | –$1 | –$340 | –$1,480 | –$4,330 |
| | Change in GDP per capita (%) | – | – | 5.2% | 0.0% | –0.8% | –3.4% | –10.0% |
| | GDP per capita growth rate | – | 1.9% | 2.4% | 1.9% | 1.8% | 1.5% | 0.8% |
| | Physical capital (U.S.$, billions) | – | – | – | – | – | – | – |
| | Change in physical capital (U.S.$, billions) | – | – | – | – | – | – | – |
| | Change in physical capital (%) | – | – | – | – | – | – | – |
| | Physical capital growth rate (average) | – | – | – | – | – | – | – |

**Table S.6—Continued**

| | | | | 2024 Scenarios | | | | |
|---|---|---|---|---|---|---|---|---|
| | | | | | Unilateral Withdrawal | | | |
| | | 2014 | 2024 Extrapolation | Two-State Solution | Coordinated | Uncoordinated | Nonviolent Resistance | Violent Uprising |
| Combined economic costs | Total GDP (U.S.$, billions) | $295 | $439 | $462 | $439 | $435 | $424 | $394 |
| | Change in GDP (U.S.$, billions) | – | – | $22.8 | –$0.2 | –$4.0 | –$15.0 | –$45.1 |
| | Change in GDP (%) | – | – | 5.2% | 0.0% | –0.9% | –3.4% | –10.3% |
| | GDP growth rate (average) | – | 4.1% | 4.6% | 4.1% | 4.0% | 3.70% | 2.90% |
| | GDP per capita (U.S.$) | $35,900 | $43,300 | $45,500 | $43,300 | $42,900 | $41,800 | $38,800 |
| | Change in GDP per capita (U.S.$) | – | – | $2,250 | –$20 | –$400 | –$1,480 | –$4,440 |
| | Change in GDP per capita (%) | – | – | 5.2% | 0.0% | –0.9% | –3.4% | –10.3% |
| | GDP per capita growth rate | – | 1.9% | 2.4% | 1.9% | 1.8% | 1.5% | 0.8% |
| | Total ten-year difference in GDP (U.S.$, billions) | – | – | $123 | –$1 | –$22 | –$80 | –$250 |
| | Total support required from international community (U.S.$, billions)[b] | – | – | $30.0 | $13.5 | $0.0 | $0.0 | $0.0 |

NOTE: Data may not match total because of rounding.

[a] All scenarios assume that any nonsecular increase in Palestinian GDP (e.g., return of refugees) is met with increased investment such that GDP per capita remains constant.

[b] For the Israelis, this includes the total ten-year costs required for supporting the removal of settlers. For the Palestinians, this includes the level of investment that will be required to satisfy the change in opportunity costs, as well as any reductions if the Israelis take over payments for PA functionality.

**Table S.7**
**Summary of the Economic Change of Present Trends for Palestine in the Five Scenarios in 2024**

| | | | | 2024 Scenarios | | | | |
|---|---|---|---|---|---|---|---|---|
| | | | | | Unilateral Withdrawal | | | |
| | | 2014 | 2024 Extrapolation | Two-State Solution | Coordinated | Uncoordinated | Nonviolent Resistance | Violent Uprising |
| Opportunity costs only | Population (millions)[a] | 4.8 | 6.5 | 7.1 | 6.5 | 6.5 | 6.5 | 6.5 |
| | Total GDP (U.S.$, billions) | $13.9 | $19.9 | $28.7 | $21.5 | $19.9 | $17.7 | $12.6 |
| | Change in GDP (U.S.$, billions) | – | – | $8.8 | $1.6 | $0.1 | –$2.2 | –$7.3 |
| | Change in GDP (%) | – | – | 44.3% | 8.1% | 0.4% | –10.8% | –36.7% |
| | GDP growth rate (average) | – | 3.6% | 7.5% | 4.5% | 3.7% | 2.5% | –1.0% |
| | GDP per capita (U.S.$) | $2,890 | $3,080 | $4,060 | $3,330 | $3,090 | $2,740 | $1,950 |
| | Change in GDP per capita (U.S.$) | – | – | $980 | $250 | $10 | –$330 | –$1,130 |
| | Change in GDP per capita (%) | – | – | 32.0% | 8.1% | 0.4% | –10.8% | –36.7% |
| | GDP per capita growth rate | – | 0.6% | 3.5% | 1.4% | 0.7% | –0.5% | –3.9% |
| | Physical capital (U.S.$, billions) | $27.9 | $38.6 | $97.0 | $48.8 | $39.0 | $38.6 | $38.6 |
| | Change in physical capital (U.S.$, billions) | – | – | $58.4 | $10.2 | $0.4 | $0.0 | $0.0 |
| | Change in physical capital (%) | – | – | 151.3% | 26.4% | 1.1% | 0.0% | 0.0% |
| | Physical capital growth rate (average) | – | 3.3% | 13.3% | 5.7% | 3.4% | 3.3% | 3.3% |

**Table S.7—Continued**

| | | | | 2024 Scenarios | | | | |
|---|---|---|---|---|---|---|---|---|
| | | | | | Unilateral Withdrawal | | | |
| | | 2014 | 2024 Extrapolation | Two-State Solution | Coordinated | Uncoordinated | Nonviolent Resistance | Violent Uprising |
| Combined economic costs | Total GDP (U.S.$, billions) | $13.9 | $19.9 | $29.6 | $21.4 | $19.8 | $17.5 | $10.8 |
| | Change in GDP (U.S.$, billions) | – | – | $9.7 | $1.5 | –$0.1 | –$2.4 | –$9.1 |
| | Change in GDP (%) | – | – | 48.8% | 7.6% | –0.5% | –12.0% | –45.7% |
| | GDP growth rate (average) | – | 3.6% | 6.9% | 4.4% | 3.6% | 2.3% | –2.5% |
| | GDP per capita (U.S.$) | $2,890 | $3,080 | $4,190 | $3,310 | $3,060 | $2,710 | $1,670 |
| | Change in GDP per capita (U.S.$) | – | – | $1,110 | $230 | –$20 | –$370 | –$1,410 |
| | Change in GDP per capita (%) | – | – | 36.1% | 7.6% | –0.5% | –12.0% | –45.7% |
| | GDP per capita growth rate | – | 0.6% | 3.8% | 1.4% | 0.6% | –0.6% | –5.3% |
| | Total ten-year difference in GDP (U.S.$, billions) | – | – | $50 | $8 | –$1 | –$12 | –$46 |
| | Required investment in public and private infrastructure by international community (U.S.$, billions)[b] | – | – | $58.4 | $10.2 | $0.4 | $0.0 | –$16.7 |

NOTE: Data may not match total because of rounding.

[a] All scenarios assume that any nonsecular increase in Palestinian GDP (e.g., return of refugees) is met with increased investment such that GDP per capita remains constant.

[b] For the Israelis, this includes the total ten-year costs required for supporting the removal of settlers. For the Palestinians, this includes the level of investment that will be required to satisfy the change in opportunity costs, as well as any reductions if the Israelis take over payments for PA functionality.

# Abbreviations

| | |
|---|---|
| BDS | boycott, divestment, and sanctions |
| GDP | gross domestic product |
| IDF | Israel Defense Forces |
| ISIS | Islamic State of Iraq and Syria (sometimes called the Islamic State of the Levant [ISIL] or the Islamic State) |
| PA | Palestinian Authority |
| PNSF | Palestinian National Security Forces |
| UN | United Nations |
| WBG | West Bank and Gaza |

# Glossary

| | |
|---|---|
| Areas A, B, and C | The three administrative areas into which the West Bank was divided by the Oslo II Accord: Area A corresponds to all major population centers, and the Palestinian Authority (PA) has responsibility for all civilian and security matters; Area B covers rural areas in which the PA has civilian control, but military matters are handled by Israel; and Area C is under the full control of Israel's military |
| Clinton Parameters | Guidelines proposed by former U.S. President Bill Clinton in 2000 |
| coordinated unilateral withdrawal | A scenario in which Israel withdraws from a good portion of the West Bank and coordinates withdrawal with both the Palestinians and the international community |
| direct costs | Specific budgetary or financial expenditures related to the conflict |
| Fatah | A major Palestinian political party, originally known as the Palestinian National Liberation Movement |
| final status accord issues | Unresolved issues under the Oslo I Accord that both parties identified as needing resolution before a final agreement was signed—including borders, refugee right of return, security, settlements, and Jerusalem |
| Geneva Initiative | A 2003 track-two draft agreement for a two-state solution (also known as the Geneva Accord); an expanded 2009 version covered final status issues |
| Green Line | Demarcation lines between Israel and neighboring territories established by 1949 armistice agreements |
| Hamas | Radical Palestinian Islamic organization |

| | |
|---|---|
| Haredi | Strictly or ultra-Orthodox Jews characterized by rejection of modern secular culture |
| Hezbollah | Shi'a Islamist militant organization based in Lebanon |
| Iron Dome | Israel's missile defense system |
| Israel | Used in this report to refer to the State of Israel and to the territory defined by the Green Line |
| Israelis | Used in this report to designate Israel's inhabitants in general; *Jewish Israelis* and *Palestinian citizens and residents of Israel* are used when appropriate to distinguish between these groups |
| Judea and Samaria | The ancient names of the kingdoms of Judea and Samaria, including all of what today is the West Bank |
| nonviolent resistance | A scenario that considers nonviolent resistance by Palestinians in pursuit of their national aspirations, including Palestinian legal efforts at the United Nations and other world bodies, continued support for trade restrictions on Israel, and nonviolent demonstrations |
| Olmert-Abbas package | 2008 negotiations between Israeli Prime Minister Ehud Olmert and Palestinian Authority President Mahmoud Abbas |
| opportunity costs | Lost opportunities for fruitful activity resulting from the conflict |
| Oslo Accords | 1993 and 1995 agreements between the State of Israel and the Palestine Liberation Organization that created the Palestinian Authority and implied a future two-state solution |
| Palestine | Used in this report (along with *West Bank and Gaza*) to refer to the area of the West Bank and the Gaza Strip as defined by the Green Line |
| Palestinian Authority (PA) | Used in this report to refer to the entity set up after the Oslo Accords to administer parts of the West Bank and Gaza |
| Palestinians | Used in this report to refer to the inhabitants (except settlers) of the West Bank, Gaza, and East Jerusalem |

| | |
|---|---|
| present trends | Our base case, which assumes that economic and security outcomes continue along their current trajectories—i.e., the final status accord issues defined in the Oslo Accords remain unresolved, and there are no significant shocks or changes to economic, demographic, or security conditions. We assume that the impasse remains dynamic, as it has always been, and that conditions, including periodic business disruptions, flare-ups of military engagement, and continued construction of Israeli settlements, continue to evolve along current trajectories. |
| Second Intifada | 2000–2005 Palestinian uprising against Israeli occupation (the First Intifada began in 1987 and ended in 1993) |
| security barrier | Israel's term for a barrier it constructed to separate Israel and the West Bank (called *the wall* by Palestinians). The route of the barrier, construction of which started in 2002, lies entirely along or within the West Bank side of the Green Line; it generally follows the Green Line, but parts of it extend into the West Bank to encompass some Israeli settlement areas. Israel's stated position is that the current barrier is not a political border. |
| State of Palestine | Used in this report when appropriate to refer to such a prospective entity, especially with respect to the two-state solution scenario |
| track-two diplomacy | Informal conflict-resolution activities conducted by private citizens (*track one* refers to official negotiations between governmental representatives) |
| two-state solution | A scenario in which a sovereign Palestinian state is established alongside the State of Israel |
| uncoordinated unilateral withdrawal | A scenario in which Israel withdraws from part of the West Bank but does not coordinate with the Palestinians or the international community, and they do not support Israel's actions |
| violent uprising | A scenario that considers the effects of a violent Palestinian uprising, perhaps emanating from Gaza but also including the West Bank and possibly participation from foreign terrorists |

| | |
|---|---|
| the wall | Palestinian term for an Israeli barrier separating Israel and the West Bank (called a *security barrier* by Israel). The route of the barrier, construction of which started in 2002, lies entirely along or within the West Bank side of the Green Line; it generally follows the Green Line, but parts of it extend into the West Bank to encompass some Israeli settlement areas. Israel's stated position is that the current barrier is not a political border. |
| West Bank and Gaza (WBG) | Used in this report (along with *Palestine*) to refer to the area of the West Bank and the Gaza Strip as defined by the Green Line |